DEADLIEST ANIMALS

AFRICAN LION

A Crabtree Branches Book

Amy Culliford

Crabtree Publishing
crabtreebooks.com

School-to-Home Support for Caregivers and Teachers

This high-interest book is designed to motivate striving students with engaging topics while building fluency, vocabulary, and an interest in reading. Here are a few questions and activities to help the reader build upon his or her comprehension skills.

Before Reading:

- *What do I think this book is about?*
- *What do I know about this topic?*
- *What do I want to learn about this topic?*
- *Why am I reading this book?*

During Reading:

- *I wonder why...*
- *I'm curious to know...*
- *How is this like something I already know?*
- *What have I learned so far?*

After Reading:

- *What was the author trying to teach me?*
- *What are some details?*
- *How did the photographs and captions help me understand more?*
- *Read the book again and look for the vocabulary words.*
- *What questions do I still have?*

Extension Activities:

- *What was your favorite part of the book? Write a paragraph on it.*
- *Draw a picture of your favorite thing you learned from the book.*

TABLE OF CONTENTS

FIERCE PREDATORS

African lions are fierce **predators** that rule over their domain. They are often called *King of the Jungle*.

Male lions, with their impressive **manes**, look like kings. Their nickname, however, is misleading. Lions don't live in the jungle. Most **inhabit** the African savannas. A smaller group of lions live in India.

A FEMALE LION, OR LIONNESS, DOES NOT GROW A MANE.

AFRICAN SAVANNAS

Savannas cover nearly half of Africa. Savannas are tropical grasslands with few trees scattered throughout. Many large animals share the savannas with lions, including wildebeest, zebras, buffalo, and rhinos. Other big cats, like cheetahs and leopards, live there too.

Lions are carnivores. They prey mostly on wildebeest, giraffe, zebra, and buffalo. Sometimes they tangle with young elephants and even crocodiles.

WILDEBEEST

DEADLY WEAPONS

WEAPON 1—TEETH

Teeth are a lion's deadliest weapon. They have four large **canine** teeth that can easily rip through skin and tear away flesh.

CANINE TEETH IN HUMANS ARE ABOUT A HALF INCH (1.5 CM) LONG. ADULT LIONS HAVE CANINE TEETH THAT ARE ABOUT 4 INCHES (10 CM) LONG.

WEAPON 2—CLAWS

Claws can prove deadly when lions use them to take down prey. Lion claws are razor-sharp. They grow about 1 1/2 inches (3.8 cm) long.

JUST LIKE THE CLAWS ON A HOUSE CAT, LION CLAWS STAY RETRACTED INSIDE A PROTECTIVE SKIN WHEN NOT IN USE.

Weapon 3—Fur Color

Lions have a natural **camouflage**—their fur. Their yellowish-brown fur blends in with the grass, making it easier to approach prey without being seen.

Lions typically **stalk**, or charge and corner, **prey**. Lions kill by clamping their jaws around their victim's neck.

Weapon 4—Keen Senses

Lions have large round eyes that are very sensitive to light—six times more sensitive than human eyes. Their pupils will open wide in the dark, allowing in moonlight. This gives lions excellent night vision.

Lions have a keen sense of smell. An organ on the roof of their mouth helps them sense nearby prey. Lions can also hear prey a mile (1.6 km) away.

LIKE MOST PREDATORS, LIONS HAVE EYES ON THE FRONT OF THEIR HEAD, RATHER THAN ON THE SIDES, LIKE THEIR PREY. THIS HELPS THE LIONS TO BETTER FOCUS ON THEIR PREY.

PRIDE LIFE

Lions live in groups called prides. Prides consist of related females, their cubs, and one or more adult males. Females care for the young, and do most of the hunting. Adult males protect the pride and its territory.

Females often give birth around the same time. They will **suckle** any of the cubs, not just their own. Females usually stay with the pride for life. Males, however, are forced out at 2 or 3 years of age. These young males often stick together as they search for a new pride.

MOM'S MILK IS NUTRITIOUS, AND HELPS THE CUBS GROW STRONG. CUBS BEGIN TO EAT MEAT AROUND TWO MONTHS OLD, BUT CONTINUE TO SUCKLE UNTIL ABOUT SIX MONTHS OF AGE.

LIONS BY THE NUMBERS

Lions are the **2nd largest** cat in the world, after the tiger.

Prides can have as many as **40 members.**

Male lions average **420 pounds** (190 kg).

Female lions average **300 pounds** (136 kg).

Pride territory can be as big as **100 square miles** (259 sq km).

Lions in the wild live about **10-15 years.**

Lions in captivity can live up to **25 years** of age.

The world's lion population is estimated to be **20,000 in the wild.**

ON THE HUNT

When lions hunt together, they spread out and stalk their prey from different directions. Lions slowly close in until they are spotted.

As their prey begins to flee, the lions begin their chase. If they catch their prey, their claws come out. A swipe of a paw across their victim's body helps to bring them down. Males take down large prey, like wildebeest or buffalo, by jumping on their backs.

LIONS ARE NOT THE FASTEST ANIMAL ON THE SAVANNA, BUT CAN REACH 50 MPH (80.5 KPH) IN SHORT BURSTS.

THE FIRST TSAVO LION TO BE TRACKED AND KILLED.

LION ATTACK!

The year 1898 was a scary time for people in the Tsavo region of Kenya. Men worked during the day building a railroad. At night, they camped in the wild African **bush.** Without warning, two large male lions began to stalk and attack the men. Some of the men were pulled from their tents, never to be seen again.

THE WORKERS' CAMPSITE

LIONS OCCASIONALLY ATTACK PEOPLE, BUT THIS BEHAVIOR IS NOT COMMON.

TSAVO LIONS ON DISPLAY IN THE FIELD MUSEUM OF NATURAL HISTORY IN CHICAGO, ILLINOIS. UNLIKE OTHER MALE LIONS, MALE TSAVO LIONS DO NOT HAVE MANES.

The Tsavo lions returned many times that year, terrorizing the workers. By the time the lions were tracked and killed, dozens of men had been taken and killed by the lions.

LIONS AND PEOPLE

Lions have more to fear from people than the other way around. Hundreds of lions die each year at the hands of humans. Some are killed because they pose a threat to humans or livestock. Unfortunately, many lions are killed for sport.

TRAVELING IN LION COUNTRY

What should you do if you come face to face with a lion? Don't run. Even the fastest human can't outrun a charging lion. A charging lion will usually turn, or stop at the last moment. If you run, you could become prey. It's best to stand your ground, facing the lion. Experts recommend carrying pepper spray to ward off an attack.

KILL OR BE KILLED

Lions and hyenas are bitter rivals. They often steal prey from each other. Lions will also kill hyenas whenever they have the chance. However, a clan of hyenas is not afraid to fight a smaller group of lions. When it comes to lions and hyenas, it's kill or be killed.

HYENAS ARE NOT AFRAID TO PROTECT THEIR KILL FROM SMALLER GROUPS OF LIONS.

Lions share the savanna with many of the world's deadliest hunters. Cheetahs, leopards, clans of hyenas, and packs of wild dogs are all fierce predators. But, for tens of thousands of years, lions have held their spot at the top of the food chain. That's why the African lion is considered one of the world's deadliest animals.

WILD DOGS

CHEETAH

LEOPARD

GLOSSARY

bush (bush): Area of land never farmed and covered with trees and bushes

camouflage (KAM-uh-flahzh): Coloring that makes animals look like their surroundings

canine (KAY-nine): Pointed teeth on each side of the upper and lower jaws

carnivores (KAR-nuh-vorz): Animals that eat meat

inhabit (in-HAB-it): Living in a place

manes (MAYNZ): The long thick hair on the heads and necks of lions or horses

predators (PRED-uh-torz): Animals that live by hunting other animals for food

prey (PRAY): An animal that is hunted by another animal for food

retracted (reh-TRACT-ed): Drawn in

stalk (stawk): To hunt or track in a quiet, secret way

suckle (SUHK-ul): Provide milk

Index

Websites to Visit

https://www.panthera.org/cat/lion

kids.sandiegozoo.org/animals/african-lion

kids.nationalgeographic.com/animals/mammals/facts/lion

About the Author

Amy Culliford

Amy Culliford has a Bachelor of Fine Arts. She has worked as a classroom drama teacher and led after-school drama programs. She avoids deadly animals of any kind.

The author would like to thank David and Patricia Armentrout for their research and help on this project.

Crabtree Publishing

crabtreebooks.com 800-387-7650

In Canada: We acknowledge the financial support of the Government of Canada through the Canada Book Fund for our publishing activities.

Produced by: Blue Door Education for Crabtree Publishing
Written by: Amy Culliford
Designed by: Jennifer Dydyk
Edited by: Tracy Nelson Maurer
Proofreader: Crystal Sikkens

Hardcover	978-1-4271-5410-1
Paperback	978-1-4271-5416-3
Ebook (pdf)	978-1-4271-5422-4
Epub	978-1-4271-5428-6
Read-along	978-1-4271-5434-7
Audio book	978-1-4271-5440-8

Printed in Canada/062024/CP20240614

Published in Canada
Crabtree Publishing
616 Welland Avenue
St. Catharines, Ontario
L2M 5V6

Published in the United States
Crabtree Publishing
347 Fifth Avenue
Suite 1402-145
New York, NY 10016

Photographs: Cover photo © The Len/Shutterstock.com, red splat on cover and throughout © Andrii Symonenko /Shutterstock.com, page 4 © RujStudio/Shutterstock.com, page 5 map © maodoltee/Shutterstock.com, female lion © Johan Swanepoel/Shutterstock.com, pages 6 and 7 background photo © Manamana/Shutterstock.com, lion with zebra photo © PREJU SURESH/Shutterstock.com, wildebeests © Photo Africa SA/Shutterstock.com, page 8 (top) © Eric Isselee/Shutterstock.com, (bottom) © Rudi Hulshof/Shutterstock.com, page 9 (top) © Mikhail Kolesnikov/Shutterstock.com, (bottom) © Villiers Steyn/Shutterstock.com, background photo © javarman/Shutterstock.com, page 10 © Marcel Brekelmans/Shutterstock.com, page 11 (top) © Alta Oosthuizen/Shutterstock.com, (bottom) © GUDKOV ANDREY/Shutterstock.com, page 12 © ert Vrey/Shutterstock.com, page 13 © Josh McPhail/Shutterstock.com, page 14 (top) © Teresa Moore/Shutterstock.com, (bottom) © Jane Rix/Shutterstock.com, page 15 (corner photo) © Chris Price at PulseFoto/Shutterstock.com, (smaller photo) © Ozkan Ozmen/Shutterstock.com, page 16 © Volodymyr Burdiak/Shutterstock.com, page 17 (large photo) © The Len/Shutterstock.com, (smaller photo) © Guido Amrein Switzerland/Shutterstock.com, page 18 © GUDKOV ANDREY/Shutterstock.com, page 19 (top) Gerrit_de_Vries/Shutterstock.com, (bottom) © Jez Bennett/Shutterstock.com, pages 20-21 b/w images public domain (source http://news.nationalgeographic.com/2017/04/man-eating-lions-teeth-kenya/), color photo museum display of 2 lions © Superx308 (Wikipedia) https://creativecommons.org/licenses/by-sa/3.0/deed.en, page 21 map © Meda01/Shutterstock.com, pages 22-23 background photo © Thammanoon Khamchalee/Shutterstock.com, page 23 (top) Stephan Dreyer/Shutterstock.com, (bottom) © Milvus80/Shutterstock.com, page 24 © Steffen Foerster/Shutterstock.com, page 25 (top) Fumanyc/Shutterstock.com, (bottom) © Rudi Hulshof/Shutterstock.com, pages 26-27 background photo © sarintra chimphoolsuk/Shutterstock.com, page 27 (top and bottom images) © MintImages/Shutterstock.com, (center) © ruek66/Shutterstock.com, page 28 © Rudi Hulshof/Shutterstock.com, page 28 (top) © Elana Erasmus/Shutterstock.com, (center) © Artem Avetisyan/Shutterstock.com, (bottom) © Donovan van Staden

Library and Archives Canada Cataloguing in Publication

Title: African lion / Amy Culliford.
Names: Culliford, Amy, 1992- author.
Description: Series statement: Deadliest animals | "A Crabtree branches book". | Includes index.
Identifiers: Canadiana (print) 20210218835 | Canadiana (ebook) 20210218843 | ISBN 9781427154101 (hardcover) | ISBN 9781427154163 (softcover) | ISBN 9781427154224 (HTML) | ISBN 9781427154286 (EPUB) | ISBN 9781427154347 (read-along ebook)
Subjects: LCSH: Lion—Juvenile literature.
Classification: LCC QL737.C23 C85 2022 | DDC j599.757—dc23

Library of Congress Cataloging-in-Publication Data

Names: Culliford, Amy, 1992- author.
Title: African lion / Amy Culliford.
Description: New York, NY : Crabtree Publishing, [2022] | Series: Deadliest animals - a Crabtree branches book | Includes index.
Identifiers: LCCN 2021021923 (print) | LCCN 2021021924 (ebook) | ISBN 9781427154101 (hardcover) | ISBN 9781427154163 (paperback) | ISBN 9781427154224 (ebook) | ISBN 9781427154286 (epub) | ISBN 9781427154347
Subjects: LCSH: Lion--Juvenile literature. | Dangerous animals--Juvenile literature.
Classification: LCC QL737.C23 C85 2022 (print) | LCC QL737.C23 (ebook) | DDC 599.757--dc23
LC record available at https://lccn.loc.gov/2021021923
LC ebook record available at https://lccn.loc.gov/2021021924